Vibrations

A. P. Sinnett

Kessinger Publishing's Rare Reprints

Thousands of Scarce and Hard-to-Find Books on These and other Subjects!

- Americana
- Ancient Mysteries
- Animals
- Anthropology
- Architecture
- Arts
- Astrology
- Bibliographies
- Biographies & Memoirs
- Body, Mind & Spirit
- Business & Investing
- Children & Young Adult
- Collectibles
- Comparative Religions
- Crafts & Hobbies
- Earth Sciences
- Education
- Ephemera
- Fiction
- Folklore
- Geography
- Health & Diet
- History
- Hobbies & Leisure
- Humor
- Illustrated Books
- Language & Culture
- Law
- Life Sciences

- Literature
- Medicine & Pharmacy
- Metaphysical
- Music
- Mystery & Crime
- Mythology
- Natural History
- Outdoor & Nature
- Philosophy
- Poetry
- Political Science
- Science
- Psychiatry & Psychology
- Reference
- Religion & Spiritualism
- Rhetoric
- Sacred Books
- Science Fiction
- Science & Technology
- Self-Help
- Social Sciences
- Symbolism
- Theatre & Drama
- Theology
- Travel & Explorations
- War & Military
- Women
- Yoga
- *Plus Much More!*

We kindly invite you to view our catalog list at:
http://www.kessinger.net

THIS ARTICLE WAS EXTRACTED FROM THE BOOK:

Transactions of the London Lodge of the Theosophical Society 1895-1913

BY THIS AUTHOR:

A. P. Sinnett

ISBN 0766131157

READ MORE ABOUT THE BOOK AT OUR WEB SITE:

http://www.kessinger.net

OR ORDER THE COMPLETE
BOOK FROM YOUR FAVORITE STORE

ISBN 0766131157

No 40.] [April, 1906.

TRANSACTIONS

OF THE

London Lodge

OF

THE THEOSOPHICAL SOCIETY

––––

VIBRATIONS

By A. P. SINNETT

PRICE ONE SHILLING

––––

LONDON
THE THEOSOPHICAL PUBLISHING SOCIETY
1906

No. 40.]   [April, 1906

TRANSACTIONS

OF THE

London Lodge

OF

THE THEOSOPHICAL SOCIETY.

VIBRATIONS

The substance of Lectures delivered to the London Lodge during the winter of 1905, by A. P. Sinnett.

Vibrations.

STUDENTS of occult science are fully alive to the significance of vibrations as associated with consciousness. As
physiology of the ordinary kind gradually attempted to trace
a connection between the brain and the states of consciousness of the being to whom the brain belonged, vague impres-
sions arose to the effect that every impression made upon
consciousness must be in some way associated with a perma-
nent change in the constitution of the brain. Memory was
supposed to represent the accumulated modifications which
the brain matter had thus undergone. This conception has
long since been dissipated for students of occultism, who
grasp the principle that the part which the brain plays in
relation to consciousness is roughly analogous to that which
the strings of a pianoforte play in reference to the music they
produce. Their vibrations provide for the manifestation of
the music on the physical plane. We need not assume that
the vibrations of the brain matter give rise to the thought,
but, in some manner as yet most imperfectly conjectured, they
are associated with the thought as affecting the physically
incarnated thinker.

At the outset of any attempt to investigate the meaning of
vibration and the manner in which this marvellous process
goes on as far as we can observe on the higher planes of
nature as well as on the physical, it is above all things important to bear in mind that consciousness has to do with a

mystery lying infinitely beyond the range of any investigation connected with the *matter* of the various planes on which consciousness may be manifest. But as far as we can observe, whenever it is manifest on any plane of matter, however refined, vibration of one sort or another goes on concurrently with its exercise.

This warning at the first glance may seem to deprive the subject of vibrations of the interest it possesses for those who hope, by studying it, to arrive a little nearer the solution of the infinite mystery in the background. But the mechanism of nature alone introduces us to mysteries which are only less unfathomable than those relating to consciousness. Physical science enables us to begin the appreciation of its complexity; occult science widens the horizon of the research; and long before we shall be justified in exhibiting impatience at the impossibility of completely comprehending the spiritual universe, we shall find the laws governing its material manifestation embracing fields of activity so vast that even the conceptions we have already been enabled to form with the help of occult developments, will assuredly appear at a later date the beginning merely of our effort to accumulate knowledge. And the study of vibrations, even though it may afford no hope of providing an exhaustive interpretation of the relations between human and divine conditions, will, nevertheless, lead us farther in the direction of comprehending some of the great principles involved in human evolution, than at the first glance we might be encouraged to anticipate.

And we are the better able to understand the nature of vibration with reference to the higher orders of matter by reason of the fact that we can begin our observation of this all-important process on the physical plane. Vibration is, of course, distinguished from irregular motion by its orderly relation to time and space. The simplest vibration we can think of is that of a pendulum. The regular swing backwards and forwards recurs in the same period of time. If we could think of the instrument as absolutely free from the retarding influences of the atmosphere and of friction, the swing once established would continue for ever like the motion of a body in space. But as we imagine the rapidity of vibrations accelerated, we soon lose touch with those of a purely mechanical order. The balance wheel in a watch, vibrating four times in a second, represents a vibration in the same order as that of the pendulum. And vibrations still of a mechanical kind may be much more rapid. The prongs of a tuning fork may be made to vibrate many thousands of times in a second, but they introduce us to a new order of vibration, that which gives rise, when imposed on the molecules of the atmosphere, to the sense of hearing as it impinges against the tympanum of the ear.

And now we find ourselves concerned with the lowest order of those vibrations which associate themselves with consciousness. We must guard ourselves indeed as students of occult science from supposing that the vibrations of the air, which it is the business of acoustics to deal with, are the only vibrations connected with the phenomenon of sound. Many people gifted with clairvoyance of a certain kind, will observe effects of colour evoked by the activity of musical instruments. And that observation concurs with broad assertions that have been put forward from time to time with varying degrees of occult authority, to the effect that whenever sound is produced by the vibration of the atmosphere, the ether which interpenetrates that atmosphere is set into vibration also, thus investing sound with much more widely ramifying influences than ordinary science attaches to the idea. But, as far as the atmospheric vibration is concerned, those which relate to sound soon reach a limit. It is difficult to draw a hard and fast line, and to affirm that beyond a certain rapidity the atmosphere cannot vibrate. But, at all events, when its vibrations exceed a rapidity in the neighbourhood of 30,000 vibrations per second, they cease to impart corresponding vibrations to the mechanism of our ears, and thus become inaudible. They can still be detected by physical plane devices. Inaudible sounds,—that is to say, atmospheric vibrations above a certain rapidity,—will affect sensitive flames, and can thus be studied in the physicist's laboratory. But the drum of the ear cannot be made to vibrate more rapidly than a limited rate—and it is well at the outset of this investigation to realise, as clearly as possible, the meaning of that state of things which plays so important a part in all considerations connected with vibrations—the inability of the various perceptive organs with which we are supplied to record vibrations in the medium to which they relate, that may be of too rapid an order.

The idea under consideration can be illustrated by a mechanical analogy. Take a pendulum at rest the length of which would provide for its vibration once per second. Imagine it tapped as lightly as you please, at intervals exactly corresponding to its own proper rate of vibration, and in time these impulses will accumulate sufficiently to produce a sensible motion in the pendulum. Of course the experiment must be carried on under imaginary conditions free from ˎatmospheric viscosity and friction, but theoretically it would work, however delicate the impulses might be, and would work in practice even though they might be very minute. But imagine these impulses recurring at a rate different from that of the rate at which the pendulum would normally vibrate, and no effect would be produced, simply because such impulses would neutralise one another rather than accumulate their effects, and thus while minute taps repeated once per second will ultimately set the pendulum in motion, taps repeated a hundred times per second will produce no effect at all. The phenomena of inaudible sound and, to go a step higher, those of invisible light, are exactly analogous to the illustration just suggested.

The interval on any scale representing rapidity, which lies between the most rapid vibrations of the atmosphere giving rise to sound, and the lowest vibrations of the ether giving rise to light, is, of course, enormous. Light vibrations begin in the orders of magnitude having to do with billions (millions of millions) of vibrations per second, and these billions are to be counted by seven or eight hundred before we reach the rapidities which the eye is incapable of perceiving. Beyond these rapidities again, if we may draw reasonable inferences from what we know of nature's higher planes, the vibrations of matter belong to those orders compared to which the ether—the vehicle of luminous vibrations—is a dense medium, represent new orders of magnitude. But before considering this reasonable assumption, incapable as yet, of course, of scientific verification, it is desirable to study the vibrations of the ether a little more elaborately.

A rapidity not greatly in excess of the highest sound vibrations introduces us to those vibrations of the ether which have to do with electric waves. Of course the phenomena of electricity are much more complicated than those which have to do with the etheric waves belonging to that group of phenomena, but these considered by themselves cover a range of vibrations beginning perhaps somewhere about the rate of a million per second up to a rapidity of over 30,000 million. That rapidity, however, is so far below the lowest rate which gives rise to the phenomena of light that we must recognise a group of vibrations between 30,000 million and about 35 billions, the effects of which must be regarded for the present as unknown. But the last named rate introduces us to the lowest heat rays of the spectrum, invisible heat rays, of course, not giving rise to the sensible condition of red heat. But spectrum rays of all sorts, from the lowest heat rays to the highest recorded ultra-violet rays, represent a range from about 35 billion to about 1,875 billion per second. To anticipate some confusions of thought let it be here remembered that according to English numeration a billion is a million million. Unfortunately continental physicists adopt a different numeration, and when they say a billion mean a thousand million. But for our purposes a billion is a million million; a trillion, a million billion, and so on.

These figures are deplorably beyond the reach of any imagination we can bring to bear upon the matter, but at the same time it is easy to understand the calculations on which they depend. The actual length of an etheric light wave is measurable by certain processes familiar to optical science, and the length of a wave of yellow light, the middle vibration of the light series, is about the 52,000th part of an inch. The velocity with which this vibration is transmitted through space is about 186,000 miles per second, and if we calculate how many times the 52,000th part of an inch will go in 186,000 miles we reach the figures just referred to.

Now for reasons connected with the mathematics of the X-ray it is assumed by modern physicists that those vibrations begin somewhere in the neighbourhood of hundreds of thousands of billions, and extend into the neighbourhood of trillions per second. Thus there is another region of vibration between the highest of the light series and the lowest of the X-ray series not yet accounted for. "We must own," writes Sir William Crookes in one essay on the subject, "our entire ignorance as to the part they play in the economy of creation "; but whether etheric vibrations belong to the lower or the higher orders of rapidity, they all seem to pass through space within about the same periods of time, the most striking characteristic of those representing the highest frequency being that they are capable of passing freely through bodies which are opaque to the vibrations of light " with scarcely any diminution of intensity, and pass almost unrefracted and unreflected along their path with the velocity of light."

Let us now go back to consider some vibrations of the physical plane, of a kind which up to the present moment we have not taken into consideration. Those which we have been considering are in all cases wave motions in great masses of the matter affected. Ripples on the surface of water radiating outwards from a splash, are waves each of which consists manifestly of enormous aggregations of the molecules of which the water consists. So when a tuning fork is set in vibration the prongs are moving as complete masses of matter. The waves of the atmosphere to which they give rise are different in their shape and character from those of the water surface, but they are waves each of which consists of molecules in enormous number. Again, when a resonant bar of metal is struck the waves rushing through its substance have a complex form that baffles imagination. But these waves are totally unlike in their character from the vibration going on all the time amongst the separate molecules of the metal. Indeed, throughout the various media in which vibrations take place, even in the ether, we have to remember that molecular vibration is going on within the wave motion that may become perceptible to the senses. This sometimes becomes manifest when excited beyond its normal rate. A difference, for instance, is perceptible between a lump of cold iron and a lump of iron heated to redness. The molecular vibration which in the cold mass was out of relation with the possible wave motions of the ether surrounding it, becomes accelerated by heat to that degree that it does eventually set the ether in motion. As the heat increases from redness to whiteness that simply means that the rate of molecular vibration in the iron has been accelerated until it corresponds with the wave length of light higher than that of the red end of the spectrum, and we even get a scientific truth roughly recognised by manufacturing metallurgists, who speak sometimes of a "blue heat" when they mean a temperature higher again than that which excites the white luminosity.

And it must be recognised that this molecular vibration is a phenomenon of matter in all its varieties, and in all these varieties quite distinct from the wave motion associated with sound, heat or light. And this thought must still be borne in mind when we come to consider those wave motions of the brain matter associated with states of consciousness, but distinctly different in their character from the molecular motion of the matter constituting the brain, which belongs in nature to a different impulse from that giving rise to what, for convenience, we may call the vibrations of thought. We do not as yet possess any data on the basis of which we can form a guess as to the order of rapidity to which those brain vibrations belong, but every convergent line of occult inquiry tends to the same conclusion, *viz.*, that the brain matter of every sentient being in a physical body is in a state of vibration whenever varying thoughts or states of consciousness are manifesting themselves in such vehicles. As the process of thought is completed for the time, the brain matter may subside into a state of rest (except for its molecular vibration, which is essential to its continued health).

Now the deeply interesting idea which this consideration

suggests has to do with what may be regarded as the gradual preparation of, or—to use a more suggestive expression—the gradual education of matter towards the purpose it has ultimately to subserve, the manifestation of consciousness on the physical plane. Let us go back in imagination to the beginnings of this vast solar system, within which our own mental activities are carried on. Occult teaching has enabled us to realise that the inconceivable Being whom we regard as the Author or God of our system, sets to work on the creation of the system in a universe already replete with matter. That matter pervades space in its most refined forms, and the molecular conditions of physical, astral, and manasic matter are due to creative activities altogether antedating the commencement of any given scheme of evolution. Physical science is moving in the direction of this idea, and the scientific world is already excited by speculations concerning the constitution of that minute corpuscle described as the " electron," with which, under another name, occult students have been familiar for many years past. For the moment the physicist, advancing along conventional roads of thought, seems leaning to regard the electron as consisting of force or energy *per se* ; some physicists would say even that it is an atom of electricity, while from the occult point of view it is an atom of physical matter in its most disintegrated condition, that in which it constitutes the ether, pervading all space ; identical as regards one of its functions with the protyle of Sir William Crookes' hypothesis, and available for the uses of Divine Beings engaged in the development of solar systems. That physicists must ultimately regard the electron as not merely a vortex of energy but as matter endowed with energy, seems reasonably certain from the point of view of occult knowledge. That statement does not controvert the metaphysical conception that the origin of matter eludes our search as we dive more and more deeply into the mysteries of Nature. But we shall not clear up one mystery by affecting to disregard the mysterious character of another, and we are not really making any significant statement if we declare that matter in its extremest refinement is merely a manifestation of spirit. It is useless for us even to attempt to think either of matter or spirit as separated one from the other. Their union is manifestation in Nature, and behind the veil of manifestation it is vain for us to attempt at present to penetrate.

Reverting now to the study of the vibrations which matter animated by spirit in all its varieties may exhibit, let us confine our attention to the atomic matter destined to provide material for the sun and planets of the physical system. This is already complex in its structure, as investigations connected with the nature of the etheric atom has shown. The atoms themselves are in vibration even within their constitution; the still finer atoms of astral matter of which they are composed are undergoing vibrations of extraordinary complexity. But these motions may be traced back to the activities of the power which calls the cosmos as a whole into manifestation. The energy of the creative power engaged in producing a solar system would appear in the first instance to be directed to the aggregation of cosmic matter into new molecular forms in which a new vibration is impressed upon the aggregates so developed. Vaguely groping after some comprehension of the relations between matter and spirit some thinkers have endeavoured to suggest that the matter which is in motion is one thing, and motion itself another. But spirit is so much more mysterious than motion, that the thought must certainly not be allowed to fetter imagination too rigidly. Still for the purposes of the elucidation now in progress we may think of the molecular vibration of the matter constituting our world as the motion specifically imposed upon it by the creative power which has engendered it from primordial conditions; and we may in that way begin to realise a process of development going on in matter as the world passes through its varying stages of growth, and tending in the direction of matter that shall be serviceable for vibrations of a higher order than those associated in the first instance with the evolution of the chemical elements.

With these alone the world is concerned during its early incandescent period, although directly we say this much we have to recognise, in passing, the fact that the incandescent periods of the world's youth are intercalated with others conducive to advancing stages of growth. But leaving all that has been said in a former transaction on " The Constitution of the Earth " to constitute a gigantic parenthesis at this period of the explanation, we may think of the gradual development of matter, after the surface of the growing world has cooled sufficiently to render the change possible, in the direction of a new order susceptible of becoming the vehicle of a new spiritual influence,—the animating principle of the vegetable kingdom. All speculation of this nature is apt to suggest a multitude of ramifications and the processes connected with what has here been called the education of matter must of course be regarded as going on concurrently with that even more wonderful process representing the evolution of consciousness in material vehicles. But of that it is impossible to deal fully while the other aspect of the vast process is especially under observation. The matter of the vegetable kingdom, it must be seen, is what may be loosely called a higher order than the matter of the mineral kingdom, and in this way it has somehow become capable of vibrating in a new way, somewhat better adapted to the higher purposes in view than the matter from which in some mysterious fashion it has emerged. Terms would hardly be used in a legitimate sense if we spoke of vegetable matter as already susceptible of vibrations identified with consciousness, but they have at all events reached upwards towards complexities compatible with the idea of life. And then in the progress of ages vegetable matter undergoes a new metamorphosis and becomes available for the uses of that more highly developed life represented in the animal kingdom. The processes of improvement, the education of the matter available as a vehicle for animal consciousness, continues steadily to advance. And although as yet no microscopic research can precisely trace the stages of that progress, we feel reasonably sure that as the animal kingdom attains higher and higher developments the susceptibility of its matter to more and more complicated vibrations is the essence of the great evolutionary change through which it is passing. Nor must we suppose that all animal matter represents in an equal degree the achievements of this evolution. The matter which builds up a human form is of varying degrees of perfection, and we must think of the matter available for brain uses as representing that which up to the present time represents the high-water mark of creative achievement as regards the physical plane.

But at once the thought will arise that brain matter itself is simply the product of organic chemistry, derived, like every other variety of matter in the human system, from the food in the first instance which is taken into it. Organic changes, however, in connection with the matter constituting vehicles of consciousness are probably more complicated changes than the first glance would suggest. Occult teaching enables us to realise that all the processes of Nature go on under the direction of elemental agency in its higher or lower forms. Few problems in occult physics are more difficult to comprehend than those associated with this state of things. Vague imagination prompts the conception of elemental agency as represented by little beings engaged so to speak in painting the flowers or building up the forms of animal life. Nature, however, works always from within, rather than from without, and the elemental consciousness, whatever form its manifestation may take under other conditions, must be thought of in regard to the progress of physical evolution, as somehow immanent in the matter itself. But however the work may be carried out, the process that has been referred to above as the education of matter will be seen to represent really what may be better described as the education of the elemental agency engaged throughout the processes of evolution in the perfection of matter. And thus, although the same material supplied by the digestive processes and the blood may be available for the different orders of elemental agencies employed, one, for example, in constructing bone matter, another in constructing brain matter, the creative ability of each kind is perfected along the lines of its own work, and thus in spite of the apparent physiological difficulty visible at the first glance there is no real embarrassment in conceiving that the brain-making elemental agency develops, as time goes on, the power of producing mechanism more and more delicately sensitive to vibrations of the higher order.

There is one variety of matter in existence which, perhaps, eclipses in perfection, at this stage of planetary growth, the brain matter of the human being, and that is the brain matter of the ant and the bee, the most wonderful speck of matter, in all probability, on this planet in the present time. But in connection with this thought a new complication arises. Occult investigation has long since shown that this world is the theatre of many different kinds of evolution, and those marvellously intelligent little creatures, the ants and the bees, belong to a category which, for various reasons, lies outside the area of animal life directly subordinate to the schemes of evolution of which the human being is the head.

And now in connection with the observation of these processes, by means of which brain matter has been brought to its present perfection,—and should ultimately attain levels of very much higher perfection,—we are in presence of the usual ontrast between the natural methods in activity during the first half of any great cyclic undertaking, and the second half, or what is commonly called the upward arc. The first half of the evolutionary process, the downward arc, represents an impulse imparted to the forms by which consciousness is manifested, due to the original impulses imparted by the creative power. A time comes, however, in connection with evolution of all kinds, when something has to be added from within to the forces which carry the evolving entity or the evolving form along the main stream of evolution. As occult students will understand by this time, the *spiritual* growth of each human being who has once attained the conditions prevalent around us now, in countries inhabited by the most advanced race as yet in existence, depends for its future course on each human being himself. That is to say, the fifth race man—acquiring, in the first instance, a comprehension of the great natural design—has to weld his own conscious energies with those behind him, as it were, to promote his further growth. He thus accomplishes the later processes of his loftier development by the exercise of those powers which he will have learned to recognise as awaiting their stimulation in his own nature. Now, with regard to the material evolution of form along the upward arc of human progress, the same rule holds good, and man himself must contribute to that further education of the appropriate elemental agency by means of which the future improvement of the vibrating mechanism, associated with human thought, has to be provided for.

The idea is subtle, and difficult to grasp with precision, liable moreover to run into conceptions which may not accurately represent the course of events. Judging by observations that have been carried out by clairvoyant observers, capable of investigations connected with the infinitely little (even a more difficult research than those which have to do with the infinitely great) they have seen reason to believe that the etheric atom itself, of which all the molecules that go to make up physical matter are built, is in process of improvement as regards its interior constitution. Constructed as it is of very numerous spiral forms, it appears that some of these are in activity and some apparently dormant. If that is so it is more than likely that Nature has somehow provided for the gradual perfection of the etheric atom, and that consequences may ultimately ensue in connection with the susceptibility of matter when this process is complete, that are likely to prove remarkable in a very high degree. But meanwhile that which we have to think of as in process of improvement as the vibratory character of brain matter becomes more and more complicated, is not the etheric atom considered apart from other manifestations in Nature, but the organic molecule itself. This molecule of course is far removed from the minutest speck that the microscope can observe, in the direction of the infinitely little, but small as it may be, the organic molecule, with the building up of which the elemental agency that we can influence is concerned, is itself composed firstly of still smaller molecules, belonging to the category of the chemical elements, and only in a secondary degree of the enormously numerous etheric atoms aggregated into the groups as chemical elements, with which each group of organic molecules is concerned. The cell is sometimes spoken of as the smallest morsel of organic matter of which we can take cognisance, but that at all events is an enormously complicated aggregation of chemical elements, and thus of etheric atoms, and it is easy to imagine that the cell is undergoing processes of improvement which do not necessarily involve any change in the individual molecules of the chemical elements in each composition,—still less any change in the substance of which the chemical elements are built. And indeed the processes of organic atom growth lie, probably, as regards the time they take, well within the vaster periods having to do with etheric atom growth. Before this world period is over the brain-building elementals will have to be guided to the construction of enormously improved machinery for the advancing egos of mankind, and yet this vast period is a very minute fraction of the great manvantaric periods in which it is conceivable that under the operation of some cosmic forces the etheric atom is undergoing improvement.

By what means can the human Ego consciously contribute to the education of the brain-building elemental agency? If he is intelligently concerned with promoting the evolution of his own spiritual consciousness he is in truth carrying out the other part of the process, however little his attention may be directed towards it. For that which is perceptible for close and qualified observation in connection with brain vibrations, introduces us to the thought that new and higher capacities of vibration become engendered within the mechanism by the activities of correspondingly elevated thought, so that even if the perfection of cerebral machinery were the only object in view, the man bent on achieving it would have to do so by processes which would incidentally be equally serviceable in advancing his spiritual welfare. The two processes of form growth and spiritual growth are in fact so intimately blended that it is very difficult in imagination to separate the one from the other.

All the the more so if we come to recognise what must be recognised if we continue this study of vibrations, the great principle that matter on the higher planes, matter, at all events, of the astral and manasic order, is in all probability undergoing processes of improvement analogous to those in activity in the physical world around us. For we should be altogether misapprehending the importance of this subject if we thought of vibrations merely in association with physical brain consciousness. We have to recognise in their characteristics the explanation first of all of what we commonly call psychic faculty, and secondly the explanation of how in most cases at the present stage of human growth, psychic faculties appear to be entirely wanting, although even in those who do not possess them consciousness may often be known to have free play on the astral, and even on the manasic envelopes of this earth. Illustrations put forward when we were considering the purely material vibrations of sound took note of the fact, that while the vibrations of one medium, the air for example, may set up vibrations of solid matter, the drum of the ear for example, when they are in the same order of magnitude, vibrations from one side may be too rapid to provoke vibrations on the other, if the orders of magnitude differ too widely. This important principle affords us a fairly complete interpretation of the phenomena, held as a rule to be so mysterious in their nature, of psychic perception. Let it be remembered first of all that the vibrations distinctly associated with consciousness are those of the brain matter and not of the external organs, the eye or the ear, which take up in the first instance the external vibrations of the atmosphere or the ether. In passing it may be worth while to point out how little real intelligence, how little appreciation of the true mechanism of consciousness, is embodied in the speculations we sometimes find in ordinary scientific writing relating to optics, with reference to the fact that the lenses of the eye must produce an inverted image of the object seen, on the retina. How does it come to pass, the materialistic thinker inquires, that in spite of this we see the objects right side upwards? Of course the answer is that it is not the retina which sees. The retina is simply part of the mechanism engaged in transmitting vibratory impulses to the brain, and it is there we must seek, although such search would be rather hopeless at the present stage of our knowledge, for the explanation of the fact that we become conscious of external objects right side uppermost.

But going back to the problems connected with psychic faculty, we have first of all to recognise that the perception of phenomena on the higher planes of Nature becomes possible for each Ego by virtue of vibrations in the matter of the astral and manasic vehicles of consciousness. These, we may suppose, on the basis of general probabilities (and adequately enlightened observation confirms the supposition) to be of a much more rapid order than those associated with the vibrations of the physical brain. The rapidity of astral vibrations, in fact, is such as to group them in a different order of magnitude from that of the vibrations appertaining to physical matter. Thus in ordinary cases they simply cannot communicate themselves to the brain, and thus, although the Ego when, as the phrase goes, " out of the body," may have full consciousness of certain thoughts or perceptions, he cannot bring them back to that aspect of his consciousness which has to do with waking physical life. That is the simple reason why the majority of people at our growth do not enjoy what is called psychic faculty. And the fact that others do, is explained with equal simplicity by the existence in their composition of an intermediate vibrating medium,—the etheric double (familiar to all the students of the septenary principles of man),—which, when constituted in an appropriate manner, is capable first of all of picking up vibrations from the astral vehicle in touch with it, and then of transmitting them with reduced rapidity to the matter of the physical brain.

We may not as yet exactly know how this process of reduction is accomplished, but familiar as we are, in music, with the complicated phenomena of overtones, which show how any given kind of vibration in the atmosphere sets up a great number of subordinate vibrations as well, it is easy to suppose that the vibrations of the etheric double, engendered by its sympathy with the astral vehicle, may engender what we may think of as "undertones,"—corresponding octaves below,—which would in their turn be commensurate in rapidity with the capacities of the physical brain.

Anyhow we now have in imagination a complete series of vibrations associated with consciousness on the highest levels in Nature which we can reach, as also with the consciousness we are exercising in our normal physical state. But at any level, however exalted, we must never allow ourselves to be victims of the delusion that consciousness and vibrations are one and the same thing. Before hoping to be in a position to understand the nature of consciousness itself we must await that future development which some time or other may enable us to understand Divinity. But the great principle to bear in mind is, that wherever consciousness is seated in any material vehicle, no matter how refined, whether that vehicle belongs to the astral, the manasic or perhaps even to some higher level, the vibrations in the refined matter which composes it, go on whenever consciousness is in activity. And just as we have recognised the superior mental capacity of the civilised man in our day as compared with that of the savage working with imperfect vibrating machinery in the third root race, as due to or concurrent with the improvement of matter, so we must also accept the great probability that corresponding processes of improvement in the matter of the higher planes enveloping this earth have been going on all the while. This conjecture may be rather startling at first sight to many people in the habit of regarding the so-called higher planes of Nature as representing conditions of perfection towards which this physical earth is slowly, blindly, struggling. But some of the most important lessons to be derived from the study of this great subject lead us in the first instance to be more respectful to the physical plane, and less mentally intoxicated, so to speak, with the wonders, from one point of view, belonging to the phenomena of planes above.

Meanwhile at the first glance conceptions which have to do with thought and consciousness appear terribly incoherent with those relating to the movement of minute particles. That movement, however minute the particles may be, however rapid their vibration amongst themselves, seems so specifically mechanical a process that one seeks in vain to unite it in imagination with the idea of consciousness. Perhaps one step in the direction of bridging the gulf, for what it is worth, it is this. Even those vibrations that come within the range of practical experiment show us some interesting facts relating to their complexity. Reference was made in the earlier portion of this paper to the fact that the wave motions in a bar of steel may sometimes be highly complicated. It has been shown by experiment that they may be so complicated in certain cases as to constrain a loose chain lying on the surface to coil itself up into a spiral. Imagination is baffled in the attempt to figure in the mind the nature of the undulations that must, in order to produce this effect, be passing through the sonorous bar. And again, when we come to examine the vibrations associated with the behaviour of a telephone we are in presence of some equally wonderful in their complexity, but in one degree more accessible to investigation. The beautifully refined methods of measurement employed by physicists at the present time in connection with the study of the telephone, enable us to realise that the actual movement of the metal diaphragm which imparts vibration to the atmospheric molecules is, to begin with, to be measured by millionths of a millimetre. But that is not all. The impulses imparted to the atmospheric molecule by any given sound are not merely impulses of a definite wave length, they are impulses, each wave of which consists of a series of minuter waves. By suitable devices these waves may be made to delineate themselves on photographic paper, with the result of showing us that definite vowel-sounds are represented by waves of the most intricate complexity, perfectly recognisable when represented by diagrams, and again subject to modifications which represent those subtle differences of sound by means of which we recognise individual voices. Thus even so relatively gross and enormous a mass of matter as the diaphragm of a telephone vibrating to produce a particular vowel-sound, say 30,000 times in a second, is so vibrating that each of the 30,000 movements is itself composed of perhaps eight or ten subordinate quivers. One has to brood over ideas of this sort, before their full significance is appreciated. And that is going on when we make use of so commonplace, physical-plane an instrument as the telephone! What is going on in respect to the vibrations set up in the brain matter of the listener who receives the telephonic message, and how are these in turn related to the messages appealing to him from a higher plane of consciousness,—so far more complicated in their character that the delicate grey matter itself breaks down in its attempt to understand them? That is the final moral of all studies connected with vibration. They enable us to realise something concerning the complexities of Nature's work, and to realise where it is wise to abandon, for the present, attempts to understand her working behind the veil.

WOMEN'S PRINTING SOCIETY LIMITED, 66 & 68, WHITCOMB STREET, W.C.

This is the end of this publication.

Any remaining blank pages are for our book binding
requirements and are blank on purpose.

To search thousands of interesting publications like this one,
please remember to visit our website at:

http://www.kessinger.net

Printed in the USA
CPSIA information can be obtained
at www.ICGtesting.com
LVHW070915231123
764443LV00076B/1438